# HOW GLOBAL WARMING & ICE AGES BEGIN & END

by
**George Sourlis**

**BS - Physics - 1964 Purdue University**
**MS- Physics - 1969 University of Arkansas**

AuthorHouse™
1663 Liberty Drive
Bloomington, IN 47403
www.authorhouse.com
Phone: 1 (833) 262-8899

Because of the dynamic nature of the Internet, any web addresses or links contained in this book may have changed
since publication and may no longer be valid. The views expressed in this work are solely those of the author and do not
necessarily reflect the views of the publisher, and the publisher hereby disclaims any responsibility for them.

Any people depicted in stock imagery provided by Getty Images are models,
and such images are being used for illustrative purposes only.
Certain stock imagery © Getty Images.

This book is printed on acid-free paper.

ISBN: 978-1-4389-4808-9 (sc)

Print information available on the last page.

Published by AuthorHouse 08/27/2020

authorHOUSE®

# CONTENTS

# INTRODUCTION

I started this book to become better informed about global warming (GW) and to clarify the subject for the public. There is just too much hype, misinformation, and incomplete information out there. All of it promotes confusion and improper interpretation.

I always thought that the primary controllers of climate were 2 large furnaces – the sun 93,000,000 miles away and the heat at the Earth's core. I still think this is the case. To this I had to add a broader view. The Earth's orbit changes from circular to slightly elliptical on a 100,000-year cycle, but the way the planet copes with the cycle still lies with internal and external heat suppliers.

The first 6 chapters deal with standard GW topics. In Chapter 1 I define "our environment", as far as this text is concerned. Chapter 2 quantifies world energy use in the 20th and 21st centuries. (A table for the 22nd century is in Appendix C.) Chapters 3, 4, and 5 relate to energy required in heating the Earth's atmosphere, water, and that required to melt all the planet's ice, respectively. Chapter 6 deals with sea level rise as ice melts. It's final level and time to reach it are discussed.

Chapter 7 diverges from ordinary GW topics. There I estimate the energy stored in the Earth's 2 metal cores, 1 solid and 1 liquid. The number is astounding – on the order of a million million milliquads (mQ).

Furthermore, in the lower mantel is stored more than in the 2 cores, and in the upper mantel there is a similar amount. This means that more than 2/3 of the internal heat is stored in the mantels. It is a lot closer to the surface than the core. Its thermal interaction with the surface is very significant.

In Chapters 8, 9, and 10 I present a different theory of global warming that makes the Earth's interior energy a major contributor. In Chapter 8 there is a 400,000-year graph showing climate history as it pertains to $CO_2$ (carbon dioxide) concentration and temperature. The new theory is strongly supported by the close connection between the two, especially during GW. Chapter 9 shows a massive increase in worldwide flooding over the last 22 years. Chapter 10 discusses more aspects of the graphs presented in chapter 8, and it connects flood data of Chapter 9 with the new theory.

Chapter 11 compares climatic numbers with very large numbers in common use. Chapter 12 addresses climate control. Chapter 13 covers some odds and ends not yet discussed, but that may be among people's concerns. Some very different ways of looking at GW concepts are stated. Chapter 14 is the conclusion, which briefly summarizes the new theory. It offers a solution to reduce greatly or stop our contribution to global warming.

Throughout this work I round off numbers (either up or down) by small percentages to make calculations and comparisons easier. In calculating and relating to these very large numbers being exact is not the point. Getting a very good approximate value to use later is what's important.

When reading ahead do not be intimidated by the immense numbers that are required to describe our environment and GW. Never focus on a single huge mind-boggling number. Always try to compare it to similar numbers to develop some feeling and perspective about whatever you are examining.

I would like to thank physicists Robert I. Keur and his son, Mike Keur, for their help, encouragement, and suggestions. I want to thank my brother, Tom Sourlis, for his insights and questions that helped along the way.

Any income over and above my direct costs will be used to support local education, environmental concerns, or toward research to develop alternative energy sources.

Chapter 1

# OUR ENVIRONMENT: IT'S SCALE

In the early 1980s interest in global warming (GW) started to grow. At the time I was a physicist working in applied research for a business machine company, and I dismissed it, because I thought that mankind's puny efforts had little chance to influence climate, which is a massive. Recently GW has great attention (and hype) in the media. It's on everybody's mind intermittently, and politicians often use it as a political football or as a photo op. Finally, tired of it all, I decided to do my own archival research and calculations.

Initial research showed that GW is loaded with huge numbers. I looked on the internet to find estimates of the volume of water in the Earth's oceans, because they can store a lot of heat and affect climate greatly. The Pacific has about 163.1 million (163.1 times 10 raised to the 6th power – 10^6) cubic miles. To most people this number is so far out of their normal experiences, that it's difficult to comprehend. Even if the number had been billions or trillions of cubic miles, most would have thought of it the same way; "Wow, that's huge!"

The Atlantic is estimated at 75.2 million cubic miles. Standing on the shore of either and looking out, each seems to stretch to infinity. Yet the numbers show they are finite. The Pacific has 2.17 times the Atlantic's volume. Both numbers are of the same order of magnitude. (2 numbers are the same order of magnitude if, when 1 is divided into the other, the result is greater than 1/10 but less than 10, depending on which is divided into which.) We relate better to the 2 large numbers by comparing them rather than thinking of each alone.

'Our environment' is a somewhat vague phrase. For most people it's really the local territory, its climate, and its infrastructure. For the purposes of this book I will define it globally as follows. Mt. Everest is less than 6 miles above sea level; the greatest ocean depths are about 6 miles. Everything within this 12-mile thick shell is our environment. There are external things that enter our environment like sunshine, meteorites, heat from the Earth's interior, etc. Once they have entered, they become a part of it. Energy is constantly entering (and leaving) our environment.

To get some sense of the scale of this environmental shell, picture yourself standing in front of a large blackboard holding a piece of chalk. Draw a circle of radius 4 feet. Now draw a horizontal diameter of 8 feet. Suppose that this is a scale drawing of the Earth, whose diameter is 8000 miles. Then each foot of the diameter represents 1000 miles, each inch about 83.3 miles, and each quarter inch about 21 miles. The width of the circumference line is probably greater than one quarter of an inch. On this scale our entire environment lies within the thickness of that outer chalk line.

In Figure 1 I have added some details of the Earth's interior structure. It is surprising. At the center is a solid sphere composed primarily of iron. Surrounding it is a thick molten liquid shell made mostly of iron. Then there is the shell of the lower mantle followed by a shell of the upper mantle. Atop all of this are the crust and, finally, our environment.

It puts things into perspective. People can sympathize more with environmentalists, and they rightly can develop some concern. Our environment looks minuscule and fragile compared to things nearby. Yet most of us know that it and we are quite resilient. Our species have been around more than 1 million years and have survived the extreme cold of many ice ages and the GWs in between.

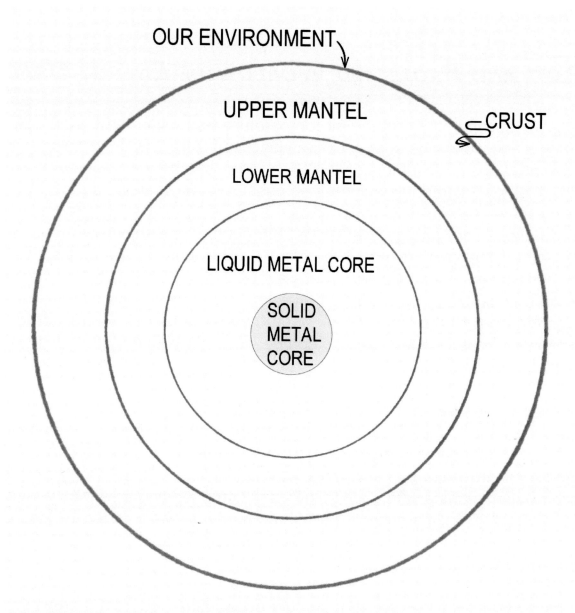

The Earth's core has 2 parts - a solid metal sphere surrounded by a thick layer of liquid metal. Both are primarily iron. Radius of the solid center, Rs, is 1220 KM. Outer radius of liquid shell, RI, is 3400 KM. Radius separating the Upper and Lower Mantels, Rul, is 4850 KM. Radius of the Earth, Re, is 6370 KM. The crust varies between 5 and 70 KM in depth below the surface.

# FIGURE 1  EARTH'S STRUCTURE

In the 20th century we experienced rising temperatures near one of those GW peaks. Ahead I hope to give more insight into our contribution to this warming trend, and how we might deal with it.

Chapter 2

# *PAST AND PROJECTED WORLDWIDE ENERGY USAGE*

It's well documented that the air temperature was rising throughout the 20th century. Since worldwide energy usage was also rising, it was natural to assume that we were and are contributing to GW. It's true, but the real question to answer is, how big is our contribution compared past GWs between ice ages?

Estimates of our energy use in the 20th century are shown year by year in texts and on the internet (see Appendix A). The sum for the tables I used is 12,829 milliquads (mQ). Whoa!! What's a mQ? It's about$10^{18}$ (10 raised to the 18th power) joules. Do you give up? Me too. What these energy units actually are isn't the point here. That it took us 100 years to use all of this energy is a benchmark for future reference and comparisons. I'll round 12,829 to 12,800 to simplify calculations later.

A starting number and an assumed percentage growth rate are needed to generate a table of projected 21st century worldwide energy use. I started with 394 mQ, the estimated usage for 2000, the last year of the 20th century. I assumed a 3% growth rate. Making this table is just like making a compound interest calculation starting with $394 in the bank compounding it year by year at 3% for 100 years. The difference is that the amounts of energy used for each year have to be added together for the hundred years to get the total. The final total is about 246,449 mQ. I'll round it 250,000 mQ (Appendix B).

That's about 20 times the 20th century usage at a modest 3% growth rate. It certainly looks like scary bad news, but it's not clear yet. Worldwide energy usage in 20th century has not yet been related to the environment.

4

# Chapter 3

# HEATING THE ATMOSPHERE

The atmosphere is a small portion of our environment, but it's the one where we measure the primary indicator of warming – its temperature. How much energy is required to raise its temperature by 1 degree Centigrade? In doing the calculation I used 5.14 x 10^18 KG as its mass. (Other estimates are usually within 10% of this value.) It takes about 1035 joules to raise the temperature of 1 KG of air at constant pressure 1 Centigrade degree. Multiplying and converting to mQ gives about 5320 mQ (Appendix D). I'll round it to 5300 mQ. Dividing the 12,800 mQ used in the 20th century by 5300 mQ per degree Centigrade gives a temperature rise of about 2.4 Centigrade degrees.

Not all of our 20th century energy warmed the air. Actual measurements show that the air only warmed about .7 degree Centigrade. The other 1.7 Centigrade degrees dissipated in the rest of our environment or was radiated into space.

The sun delivers 15,000 mQ daily to Earth. So our 20th century usage is minuscule compared to the amount the sun delivers in a century. (Always try to think in terms of comparisons.)

What about the energy we will be using in this century? Projections show it to be about 20 times what we used in the 20th century. Does that mean we can expect 20 x .7 Centigrade degrees or a 14 Centigrade degree increase in air temperature? Not really. It is unreasonable to expect this much of an increase, because we have already lived through about 7% of this century, and an increase of only .15 Centigrade degrees has been measured. At that rate what we might expect 2.1 Centigrade by the year 2100. (See Appendices B & C for 21st & 22nd century energy usage details.)

Chapter 4

# *Earth's Water - A Massive Heat Absorber*

Earth's water has a mass of about 1.36 X 10^ 21 KG. It takes about 4180 joules to raise the temperature of 1 KG of water 1 Centigrade degree. Multiplying the two and converting joules to mQ shows that it takes 5,680,000 mQ to raise all of the planet's water by 1 Centigrade degree. (Appendix E) That's 5.68 million mQ!

The projection for the 21st century usage is about 250,000 mQ. Dividing 250,000 mQ into 5,680,000 mQ shows that it takes almost 23 times this amount to raise all of the water 1 Centigrade degree! 250,000 mQ can raise the ocean temperature by only .044 Centigrade degrees. That's probably too small to be measured with any confidence. Any data that shows otherwise might be a mis-interpretation of a local anomaly. Note also that much of the 250,000 mQ will be dispersed elsewhere.

Chapter 5

# THE EARTH'S ICE

Estimates of the Earth's ice volume range from 30 to 33 million KM^3. Calculations using the higher number show that melting it will take 10 million mQ (Appendix F). Less than 5,000,000 mQ will be used worldwide in the 20th, 21st, and 22nd centuries combined. That could melt less than half of the ice. By about 2225 we will have used enough energy to melt it. Even then all is not likely to melt; much of the energy will be dispersed elsewhere.

About 5% of Earth's ice is afloat. Most is in the Arctic, some is in the ice packs around Antarctica, and some is in icebergs. When it melts, contraction fits it exactly into the space originally displaced (Archimedes Principle). None of this ice will cause sea level to rise, and there are at least 3 benefits to its melting. One is no more icebergs; ships will no longer face this threat. Another is the opening up of the famed Northwest Passage. That shortens the trip around the Arctic ice by thousands of miles. Finally, the Arctic offshore oil is now accessible.

Another 5% of the ice is in land-based glaciers. Any land-based ice can melt and run off via rivers into the oceans increasing sea level.

It is argued that glaciers are a major source of fresh water for some large cities as well as many towns and villages. The loss of glacial water, they say, will be a "catastrophe" for many of these cities. Ridiculous. It is doubtful that any of the larger cities will disappear because their primary water source disappears. To replace losses they can drill for it; some near the oceans can set up desalinization plants; others can build pipelines to bring it in, as many do for their oil and natural gas supplies. Towns and villages can do likewise.

Chapter 6

# *The Rise in Sea Level*

There are 3 large ice sheets that contain the other 90% of Earth's ice; one is in Greenland and two are in Antarctica. These, when melted, can raise sea level. Their melt water will be 90% of 30 million KM^3 (Appendix F) = 27 million KM^3.

Any volume can be expressed as an area times a height (volume = area x height). Above is the volume, which may flow into the oceans, which cover 70% Earth's area. Dividing that area into volume gives the increase (height) in sea level, 74.8 meters (Appendix G).

This number comes from the most fundamental numerical approach. It is a little too high. The masses of ice in the Antarctic and in Greenland are so large that they compress the Earth's crust below, lowering it. A sizable fraction of ice is below sea level. It will not contribute to increasing the sea level.

In some places measurements of ice depths show them to be 1 KM or more below sea level. It's hard to say what that volume is; a reasonable guess might be between 10% and 25% of the total. This approach reduces the estimated rise a bit. The range becomes about 56.1 to 67.3 meters.

Consider the time involved to get this rise. Calculations In the last chapter show that it will take about 217 years to use enough energy to melt the ice. That gives a maximum rise of .315 meter per year – 31.5 cm/year or about 1.03 feet per year. It's easy to respond to this very slow rise. It is no serious threat to life.

# SUMMARY 1  CALCULATIONS & CURRENT BENCHMARKS

## WORLD ENERGY CONSUMPTION (mQ)

| | | |
|---|---|---|
| 20th Century | 12,800 | Appendix A |
| 21st Century | 250,000 | Appendix B |
| 22nd Century | 4,700,000 | Appendix C |

It takes about 5320 mQ to warm the air (mass = 5.41 x 10*18 kg) 1 Centigrade degree.

It takes about 5,690,000 mQ to raise the temperature of Earth's oceans (mass = 1.36 x 10^21 kg ) 1 Centigrade degree!

It takes about 23 times the amount of energy we will use in the 21st century to raise the temperature of the Earth's water 1 Centigrade degree.

The oceans' mass on Earth is about 3980 times that of the atmosphere. Its higher heat capacity gives it tremendous thermal inertia.

Estimates of the volume of ice on Earth range from 30 to 33 million KM^3. It requires about 10 million mQ to melt it. By about the year 2225 civilization will have cumulatively used this much energy. It is doubtful that the ice will melt by then, because the energy we use is not dedicated to the melting of ice alone.

If all Earth's ice melts, sea level will rise in the range of 56.1 to 67.3 meters. It will take more than 2 centuries, if our energy usage is to produce this result. This rise is equivalent to about 1.03 feet per year. It is not a serious threat, but rather an inconvenience that requires a response. Yes, it will cost to relocate. The cost of living ahead will rise, and that is nothing new.

Chapter 7

# THE EARTH'S INTERIOR

The sun and the Earth's interior are sources of thermal energy that govern our environment's climate. GW scientists talk about the sun and the greenhouse effect, but they never mention the energy stored in the Earth's interior. Calculations in Appendix H show that at least $3.62 \times 10^{12}$ mQ is stored there. That's 3.62 million million mQ! The temperature of the core is not known exactly, but is thought to be in a range from 5000 to 7000 degrees Centigrade. The table in Appendix H shows energies stored there for this range in increments of 500 Centigrade degrees.

What's more, the lower and upper mantels together contain about 2.5 times the energy in the core, because most of the Earth's mass at temperatures higher than ambient (15 degrees C) is in the 2 mantels. This means that most of the interior energy is a lot closer to the surface than 3400 KM (2113 miles).

In the March 1989 issue of Scientific American magazine was published an article titled "Modeling the Earth's Geochemical Carbon Cycle". Part of that cycle has to do with tectonic plate movement. As plates move, portions of the sea floor subduct under continental shelves. The sea floor is loaded with natural carbonates as well as millions of years of accumulated carbonates (sea-life shells and bones). As it subducts, it is heated by friction and by surrounding hotter materials beneath the crust. The intense heat breaks down the carbonates releasing $CO_2$ into the atmosphere through volcanic action, soda springs, etc. The more the continents move and sea floor subducts, the more both are released into our tiny environmental shell.

Chapter 8

# A DIFFERENT THEORY OF GW

In Figure 2 are 2 graphs showing 400,000 years of temperature and CO2 data. One graph shows temperature in degrees Centigrade; the other CO2 concentration in PPM (parts per million). These data come from ice cores taken by the Russians at Vostok, Antarctica. The temperature graph (blue trace) has a range from −10 degrees to +4 degrees Centigrade. It is designated "temperature change from present". The range for CO2 concentration (red trace) is from 160 PPM to 300 PPM. (Current internet sources show concentrations of about 360 PPM.) On the horizontal time scale each space is 10,000 years. It is so compressed that 20th century detail is invisible.

The composite in Figure 3 is an overlay of the 2 graphs in Figure 2 (red over blue). This shows the close correlation between temperature and CO2 concentration; it's especially tight on the 4 sharp rises in temperature. These used to be called inter-ice-age warming periods. The one just experienced is called global warming (GW). This is the first time in recorded history that we have seen this. Hence the interest, excitement, and worry about it.

Focus on the temperature graph (blue trace) in Figure 2. At least 2 kinds of thermal cycles are obvious. There are 4 GW periods. They are the smooth fast-rising lines. Each pair is separated by a long term cooling cycle (LTCC) – an ice age. GWs last about 10,000 years, while LTCCs cover roughly 90,000 years. Warming is very efficient compared to cooling.

The GW-LTCC 100,000-year cycle was first attributed to changes in the Earth's orbit over that time from circular to slightly elliptical. Mathematician, Milutin Milankovitch, derived the equations explaining this periodicity.

FIGURE 2

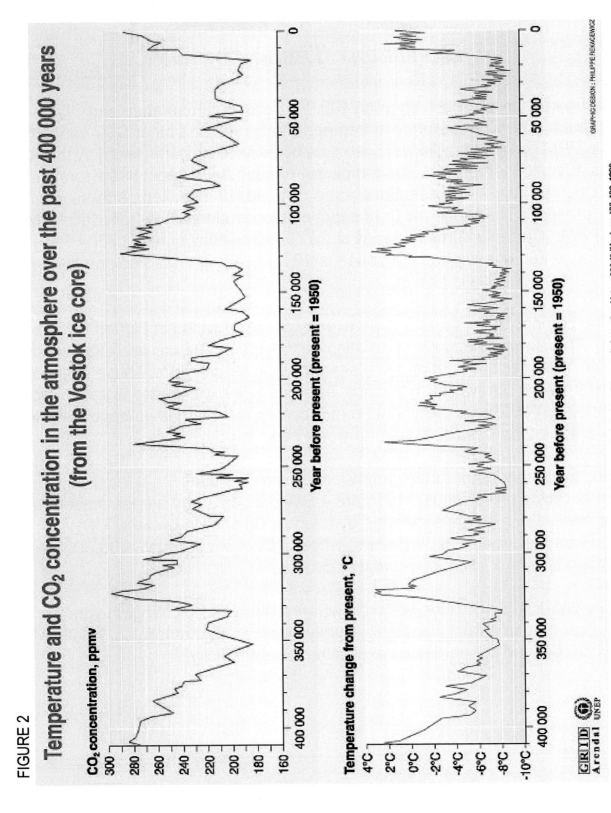

Temperature and CO₂ concentration in the atmosphere over the past 400 000 years
(from the Vostok ice core)

FIGURE 3

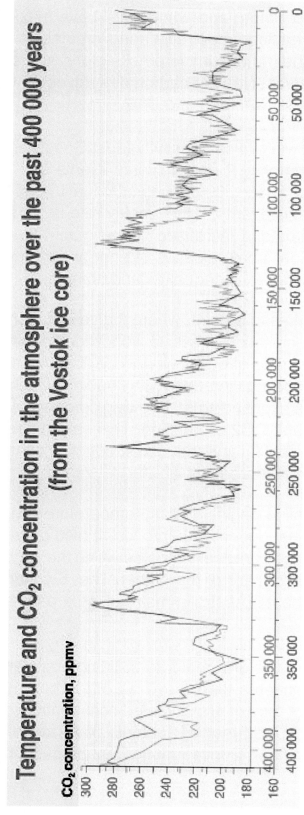

Temperature and CO$_2$ concentration in the atmosphere over the past 400 000 years
(from the Vostok ice core)

I have overlaid the CO2 graph onto the temperature graph to show how well these two correlate, especially during the GWs..

Temperature and CO2 concentration in the atmosphere over the past 400 000 years." UNEP/GRID-Arendal Maps and Graphics Library. 2000. UNEP/GRID-Arendal. 19 Jul 2008 <http://maps.grida.no/go/graphic/temperature-and-co2-concentration-in-the-atmosphere-over-the-past-400-000-years>.

Cartographer/Designer:
        Philippe Rekacewicz, UNEP/GRID-Arendal

During the 4 GWs snow and ice coverage are decreasing to a minimum near and beyond the peak, and the tectonic plates (and continents) have maximum freedom of motion. Subduction of the sea floor under continents creates heat via friction. Sea floors covered with centuries of accumulated skeletons and shells contain a high concentration of carbonates. Heating carbonates drives off $CO_2$. This gives a very tight correlation between warming and $CO_2$.

The Smithsonian has a web site showing volcano activity (Just Google Smithsonian volcano.). Weekly they are now noting 2 new volcanic events and 12 to 14 repeats. Average is about 15 listings. That's 780 per year. Over a 10,000-year GW period the total is 7,800,000 volcanic events, if the weekly rate is constant. It's the greater velocity of tectonic plates and increased volcanic activity along plate boundaries during a GW, that allow small releases of energy spread both geographically and temporally via volcanoes, earthquakes and tsunamis.

When the temperature during GW rises to a point, where it gives evaporation large enough to produce excessive precipitation, massive worldwide flooding increases with time.(See Chapter 9 for records showing worldwide flood increases from 1985 through 2007.) Evaporative cooling increases with flooding, which also increases water mass inside and on the exposed crust. Growing continental inertia slows subduction, slowing heating and $CO_2$ production. This culminates in a GW turnaround. Afterward evaporative cooling and reduced subduction continue. Both cool, while the later also reduces $CO_2$ production.

After a GW peak, a 90.000-year cooling begins. Falling temperatures change rain to snow. It piles up compressing that underneath into ice. It also grows the southern and northern ice caps. Ice bridges gradually form between the caps and the continents, until each cap covers about 1/3 of the planet. The 6 continents and 2 poles become a single united mass. (Australia may not be a part of this collective.)

The caps are like pieces of a 3-D puzzle or a mold, form-fitting precisely the contours of the continents. The 6 continents then move as one. The coupled complex has greater inertia. Millions of tons of accumulated continental ice and snow, that was not present during GW, add inertia. Tectonic plate motion decreases, as does subduction of the sea floor. Consequently $CO_2$ production and heat generation decrease.

The character of a LTCC is quite different from the smooth quick rise of a GW. There are many erratic sudden (on the time scale used here) downward dips in temperature. All are the result of volcanic eruptions, which cause another type of thermal event, a volcano-induced cooling cycle (VICC). Each major eruption puts so much ash into the air that the sunlight is wholly or partially blocked for many months, perhaps even years, at the start of the VICC. The temperature drops. Gravity and precipitation remove most of the ash over a long period after the eruption. Snow accumulations for a few thousand years reflect much sunlight giving more cooling. In the last half of the VICC sunlight, tectonic plate movement, and heat from the interior, restore conditions to some warmer equilibrium point.

Several megavolcanic eruptions can be identified on the CO2 graph (red trace) in Figure 2. The large "V"-shaped depression between 55,000 and 75,000 years ago (YA) was caused by the megavolcano at Lake Toba, the one featured in the 2007 PBS Nova program, "Mystery of the Megavolcano". Centered on about 250,000 YA are two megavolcanic eruptions at Lake Taupo, New Zealand. The two eruptions in the range of 330,000 to 350,000 YA were probably from the ring of fire around Lake Taupo. Finally, the massive eruption centered on 225,000 may be the eruption at Mauna Kea. These last 2 identifications are not certain. Megavolcanoes give the longest VICCs lasting thousands of years.

Megavolcanoes and other large volcanoes have a dual nature. Thermal overshoots afterward release heat and CO2 into the environment. Nowhere is this shown better than right after the largest megavolcano , Mauna Kea, eruption. (It may be that a magnetic reversal in the core at about this time enhances this lengthy unusual temperature rise.) Afterward from 222,000 to 200,000 years ago there is a double-peak heating event (blue trace). It's far above the "normal" thermal track that "should" have followed the GW peak.

Thermal overshoots are common after major volcanic eruptions.  Each overshoot is a "mini-GW".  Each is the result of the eruption's shaking up the tectonic plate boundaries and breaking many ice bridges. Inspect the blue trace a little more, and others become apparent.

The mechanism for volcanic eruptions during a VICC is quite different from that during GWs. Great areas of the crust are covered by millions of tons of ice and snow. Volcanoes are not excluded from this coverage. Many develop snow-ice "plugs" above the calderas. Some that are warmer develop huge lakes.  In other cases lakes develop between the crust and ice plugs. In all cases much more energy is required to erupt and remove the increased mass.

As thermal energy from the magma source continues to arrive, the reinforced heavily-loaded crust above these calderas allow them to grow. A critical point is reached, when the energy and pressure in the caldera approach the strength required to overcome its burden above. Earthquakes become more frequent. Some make cracks in the crust. Small cracks may allow the magma to reach the water above. These self-heal as the magma solidifies. Larger cracks can inject so much magma into the water that massive amounts are vaporized.

When 1 gallon of water is vaporized at atmospheric pressure, it becomes more than 22,000 gallons of vapor. If most of the water in a confined lake vaporizes and tries to occupy the same space as its water originally had, the terrific pressure can give an immense explosion (eruption).

Once started, the snow falls continually during the LTCCs. On continents millions of tons compress the ice near the ice-crust interface, driving the air out of it. The crust is also strongly compressed by the weight. Compression increases the thermal conductivity of both. The temperature differential between the crust and the interior has increased by more than 20 Fahrenheit degrees. The temperature difference between the snow and the interface is small. Snow insulates. These effects increase heat flow from the interior to the ice-crust interface.

More ice melts, but the temperature does not change there. The extreme pressure aids melting as it squeezes water away from the interface and keeps the ice in contact with the crust. This flowing water under high pressure erodes the ice. Many feet of snow and air-filled ice above the interface are such good insulators, that they prevent significant heat loss.

Simultaneously, the heat migrates laterally to uncovered areas at edges of the ice caps. At some point there is enough heat arriving, and so little ice and snow accumulating, that the growth of the ice caps stops. The perimeters begin to retreat. The melt down has begun. It is sustained by heat from the interior combined with high thermal conductivity in the newly exposed compressed crust.

As the ice caps shrink, the ice bridges are melting and weakening. If allowed to melt breaking the connection, tectonic plates would break free, increasing sea floor subduction. But large volcanic eruptions are frequent enough, that one usually intervenes, and begins the GW rise first. This can be seen at the base of all 4 GW events. Each was started via a sizable volcanic eruption.

The boom generated by each huge eruption (double booms at Taupo), which freed up the tectonic plates, was probably thousands of times louder than the loudest sound heard at the 1980 Mt. St. Helen's eruption. The sonic pulse swept through the crust around the world breaking ice bridges. It probably swept through the interior also. So some tectonic plate boundaries may have experienced more than 1 shakeup, as huge sound pulses through the crust and the interior arrived at different times.

Freeing the tectonic plates ends a LTCC and begins GW.

Chapter 9

# *Floods: More About the End of the Current GW*

In 1985 Dartmouth College began tracking worldwide floods. A table in Appendix J shows annual numbers. Over the last 22 years there has been a dynamic increase. Below is a short summary.

| YEAR | NUMBER OF FLOODS WORLDWIDE |
|------|----------------------------|
| 1985 | 69 |
| 1990 | 104 |
| 1995 | 112 |
| 2000 | 102 |
| 2005 | 171 |
| 2006 | 232 |
| 2007 | 244 |

The number of floods has increased from 69 in 1985 to 244 in 2007. Continental mass has been increasing, which slows the tectonic plates and continents, and GW ends. See complete annual records in Appendix J.

The Dartmouth Flood Observatory web site at http://www.darmouth.edu%7Efloods/Archives/index.html has great worldwide maps showing annual flooding for each year from 1985 through 2007 as well as the annual numbers shown here.

# Chapter 10

# MORE ABOUT THE *400,000*-YEAR GRAPHS IN FIGURE 2

The graphs in Figure 2 are a geoclimatic fingerprint (or DNA if you prefer) of the last 400,000 years of climate. Though much information has already been extracted to support the concept, that the sources of global warming are tectonic plate movement, subduction of the sea floor, heat from Earth's interior, and the action of volcanoes, there is more to learn.

From the quick-rising slopes of the 4 GW events the amount of time for the atmosphere to gain 1 Centigrade degree can be calculated ( Appendix K). The slopes visually appear to be just about the same, and they are quit close for this kind of calculation. Average is 306.5 years per Centigrade degree. The difference between the highest and lowest is 136 years per Centigrade degree.

| GW # | Approx. Years Between 1 Degree C Rises |
|------|----------------------------------------|
| 1 | 374 (oldest) |
| 2 | 238 |
| 3 | 372 |
| 4 | 242 ("current") |

Compare the above with the .7 Centigrade degree rise recorded in the 20th century. That's about 1 Centigrade degree every 142 years. It is not a part of 4th GW rise, and is too short a time period to support any long-term conclusion. The only one that may be valid is that it does not represent the same thermal process as GW.

The lengths of the ice ages appear to be increasing. Interesting, but I'd like a larger sample to conclude that ice ages are getting longer.

The 2 graphs in Figure 2 end in 1950. Compare the red and blue traces on the right (most current). The blue temperature trace oscillates while moving horizontally; the $CO_2$ (red trace) concentration continues to rise

In the 50+ years not shown on these graphs, others show the temperature still oscillating moving horizontally, while the CO2 concentration has skyrocketed to from 270 PPM to 360 PPM currently – a 33% increase. That's 6 full spaces above the top of this graph. The separation, which began about 3000 years ago, is immense compared to any other in Figure 3. So the present GW is distinctly different from the others.

The separation indicates that GW and CO2 are not as closely connected as was once thought. (See the excellent Vostok CO2 vs. temperature graph at the Koshland Science Museum on the internet to view the missing years.)

The quickness of the turnaround from heating to cooling is supported by the sharpness at the peaks of the previous 3 GWs in Figure 2. The 2 sharpest lasted 1000 years or a little less.

In the smooth downward slopes immediately after turnarounds at the peaks of the first 2 GW periods, there is no obvious evidence of major volcanic eruptions. Apparently it was a period of numerous small ones, but, since the CO2 concentration reduction tracked almost exactly the dropping temperature, subduction of the sea floor under the continents was decreasing due to increased continental water (inertia).

The eruption at Lake Toba 75,000 YA was featured in the PBS Nova program "Mystery of the Megavolcano". Analysis of sea floor cores showed that sea water temperature dropped about 5 Centigrade degrees over a period of about 5000 years. Since it takes about 5,700,000 mQ to change the ocean temperature 1 Centigrade degree, the oceans lost about 28,500,000 mQ in 5000 years – 50 centuries. This is 570,000 mQ per century.

This eruption, that might have taken less than a week (or a month) to complete, caused the oceans to lose 28,500,000 mQ over a period of about 5000 years. The sum of world energy usage for the 20th and 21st centuries is about half 570,000 mQ. Our energy contributions have been small compared to Mother Nature's, which are often huge.

# 100,000 YEAR CLIMATE CYCLE ON EARTH

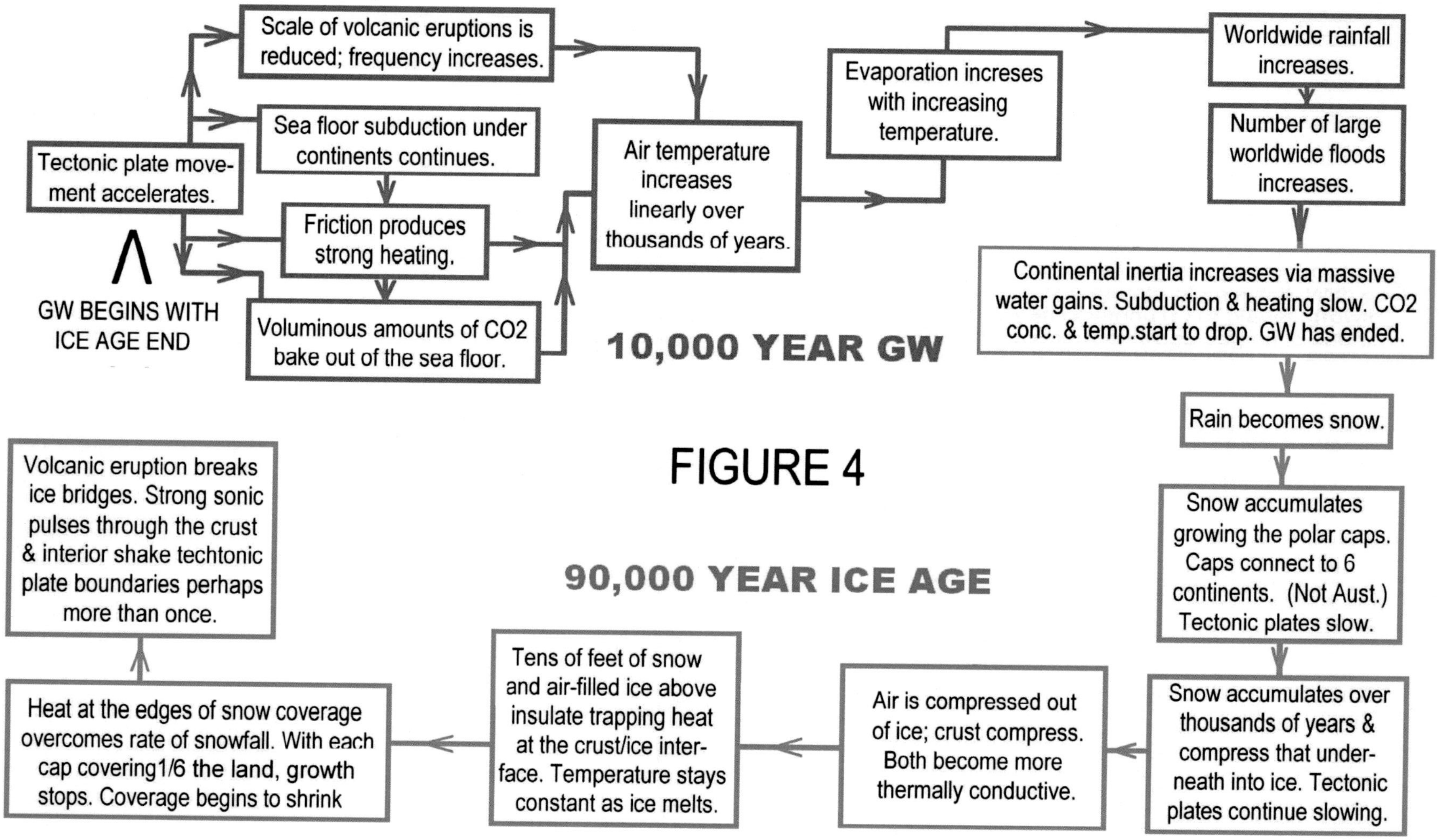

Worldwide rainfall increases.

Number of large worldwide floods increases.

Continental inertia increses via massive water gains. Subduction & heating slow. CO2 conc. & temp.start to drop. GW has ended.

Rain becomes snow.

Snow accumulates growing the polar caps. Caps connect to 6 continents. (Not Aust.) Tectonic plates slow.

Snow accumulates over thousands of years & compress that under-neath into ice. Tectonic plates continue slowing.

Evaporation increses with increasing temperature.

Air temperature increases linearly over thousands of years.

**10,000 YEAR GW**

**FIGURE 4**

**90,000 YEAR ICE AGE**

Air is compressed out of ice; crust compress. Both become more thermally conductive.

Scale of volcanic eruptions is reduced; frequency increases.

Sea floor subduction under continents continues.

Friction produces strong heating.

Voluminous amounts of CO2 bake out of the sea floor.

Tectonic plate move-ment accelerates.

∧

GW BEGINS WITH ICE AGE END

Tens of feet of snow and air-filled ice above insulate trapping heat at the crust/ice inter-face. Temperature stays constant as ice melts.

Volcanic eruption breaks ice bridges. Strong sonic pulses through the crust & interior shake techtonic plate boundaries perhaps more than once.

Heat at the edges of snow coverage overcomes rate of snowfall. With each cap covering1/6 the land, growth stops. Coverage begins to shrink

# Summary 2 More Benchmarks and Conclusions

In the last 400,000 years there have been 4 GW periods. They are natural periodic events that occur between ice ages. Civilization did not significantly contribute to the first 3 (Polar bears survived them.).

During a GW the continents and tectonic plates move more freely than in a LTCC. Sea floors subduct beneath continents. This movement and subduction generate lots of heat and $CO_2$. The heat from the friction of subduction, and other heat released from the interior are the primary causes of GW.

GWs end when exposed crust accumulates enough water mass to slow them. Slowing reduces subduction. Heat and $CO_2$ production drop.

Warm-up time during GW is much shorter than the cool down afterward (LTCC). This indicates 2 different thermal processes.

"Fast" moving tectonic plates having numerous frequent "small" volcanic eruptions along boundaries release energy continually in small bursts. During a LTCC massive eruptions attributable to increased mass and/or ice-snow plugs above calderas release MACRO-busts of energy. It is likely that this mechanism prolongs ice ages.

The heat stored in the Earth's core is on the order of $10^{12}$ mQ. More than twice as much heat is stored in the 2 mantels. So more than 2/3 the interior heat is a lot closer to the surface than the cores. It cannot be ignored in any GW theory.

GW is neither an extinction event, nor is it catastrophic for mankind. The reason is that it is happening so slowly. We survived previous GWs without as much technology, knowledge, and population as now. Indications are that we have more than 2 centuries to adapt to the LTCC after this GW.

$CO_2$ and temperature increases "currently" are no longer correlating closely. We are now in a cooling cycle. These 2 facts indicate that $CO_2$ and the greenhouse effect are not capable of overriding the natural LTCC now beginning. (More on this in Chapter 13)

Current record rainfalls and flooding are part of the LTCC after GW. It is unlikely they will abate within a lifetime. Expect a lot more very bad dangerous weather.

# Chapter 11

# WORLD CLASS NUMBERS

I define world class numbers (WCN) as those needed to describe the MACRO-physical properties of a planet. Table 1 below shows WCN. Below that is Table 2, which shows other more common numbers (OMCN). After reviewing the tables take an eye-opening look at some comparisons.

## TABLE 1  WORLD CLASS NUMBERS (WCN)

| | |
|---|---|
| Energy stored in the Earth's Iron Cores | 3 - 5 x 10^12 mQ |
| Energy Stored in Upper & Lower Mantels | 7 - 10 x 10^12 mQ |
| Daily Energy Received from Sun | 15,000 mQ |
| Mass of the Earth | 5.98 x 10^24 KG |
| Mass of Water on Earth | 1.31 x 10^21 KG |
| Mass of Air on Earth | 5.18 x 10^18 KG |
| Area of Earth | 1.97 x !0^8 MI^2 =5.1 x 10^8 KM^2 |
| Volume of Ice on Earth | 33 x 10^6 KM^3 |
| Energy Required to Melt All of the Ice | 10^7 mQ |
| Energy to Raise Earth's Water 1 Deg. C | 5,680,480 mQ |
| Time For 1 Milankovitch Cycle | 100,000 Years |
| | 1000 Centuries |
| Typical Time for GW to Begin & Peak Out | 10,000 Years |
| | 100 Centuries |

## TABLE 2  OTHER MORE COMMON NUMBERS (OMCN)

| | |
|---|---|
| 20th Century World Energy Consumption | 12,800 mQ |
| 21st Century World Energy Consumption | 244,000 mQ |
| 22nd Century World Energy Consumption | 4,700,000 mQ |
| Number of People on Earth | 6.6 x 10^9 |
| Number of pairs of People on Earth | 3.3 x 10^9 |
| Mass of a Large Person | 100 KG (Weight = 220 LBS) |
| Mass of World Pop. @ 100 KG Per Person | 6.6 x 10^11 KG |
| Life Span of a Person | About 77 Years |
| Area of Texas | 261,820 MI^2 = |
| | 677,820 KM^2= 7.2991 X 10^12 FT^2 |

## POPULATION MASS VS. MASS PROPERTIES OF EARTH & SOME OF ITS PARTS

The mass of Earth (5.98 x 10^24 KG) is 9.06 x 10^12 times as massive as the planet's population (.66 X 10^12 KG) assuming 100 KG per person. That's 9.06 million million times the mass of the population! Mass of the water (1.31 x 10^21 KG) is about the 2 billion times as massive as the world population. The mass of the air (5.18 x 10^18 KG) is about 7.8 million times as massive.

## THE AREA OF TEXAS VS. NUMBER OF PAIRS OF PEOPLE

Dividing the area of Texas in square feet (7.2991 x 10^12 FT^2) by the number of pairs in the world (3.3 X 10^9) shows that there are about 2212 FT^2 per pair available. Amazing! There's no need to worry that rising sea levels will not leave enough living space for all of us. Agricultural land to support them is addressed in Chapter 13. There will be more than we need.

## LIFE SPAN VS. THE MILANKOVITCH CYCLE OR G.W.

The roughly 100,000 year Milankovitch Cycle, 1000 centuries, is about 1299 times the current average life span now (77 years). GW periods are roughly 100 centuries vs. roughly ¾ century. More accurately that's about 130 times life span.

## AREA OF EARTH VS. AREA OF TEXAS

The area of Earth (1.98 x 10^8 MI^2) is 752 times the area of Texas (261,820 MI^2). Texas is really large, isn't it?

## ENERGY FROM THE SUN VS. 20TH CENTURY WORLDWIDE USAGE

15,000 mQ is the energy that the sun shines on the planet in 1 day, but about 20% is reflected by snow and ice. So only about 12,000 mQ is useful daily. World-wide energy usage in the 20th century was about 12,800 mQ. It took us a century to use the amount of useful energy that the sun delivers daily.

## AMOUNT OF ENERGY AND TIME TO MELT ALL THE ICE

The amount of energy required to melt all of the planet's ice is about 10,000,000 mQ. It will take the addition of worldwide energy use in the 20th, the 21st, the 22nd, and 25 years into the 23rd century before that number is reached, assuming a 3% growth rate. That's about 217 years from now.

Above are some of the ways to think about our society vs. GW and climate. The two Texas comparisons are excellent examples of surprises that show up.

Mankind and our whole infrastructure were present and grew during this current GW. But our input to it was nominal, and we certainly did not initiate it. Megavolcanic or very large eruptions initiated each of the last 4 GWs (See Fig. 2).

In fact, this last GW highly aided the growth of civilization. Hundreds of thousands, if not millions, of "small" but frequent volcanic eruptions and earthquakes, had little effect when compared to the megavolcanic activity during the LTCCs. There were no megavolcanic eruptions, which ejected materials that would have blocked the sunlight for months, severely hindering food supply, killing thousands during eruptions, and starving many more afterward. Don't forget that some of these were extinction events for some species.

During the LTCCs it was a struggle to stay alive. The dark and extreme cold after VICCs were survived by just a small percentage of the population. Continual searches for food and migrations to get more livable conditions were at the top of to-do lists. Progress like that experienced in the last few thousand years would have been impossible.

Chapter 12

# CLIMATE CONTROL

Mankind has always been egocentric. Before Copernicus the prevailing view was that the heavens revolved around Earth. That's been cleared up, but there's a new gross misconception out there. Climate control is the current buzz.

Nonsense! Climate control is now only possible in enclosed spaces. It's commonly done in houses, factories, and larger enclosures. Heating in winter and air conditioning in summer are mechanisms used.

Outdoors is another story. The best that can be done is that air pollution in large cities can be reduced significantly by limiting automotive emissions. Heavy industry can be rezoned, and pollution controls can be installed. New improved processes can be implemented at the pollution cites.

Beyond that climate control in the open is a fantasy. Not the technology, not the know-how, and not the energy supply needed to cope with Mother Nature are available. Droughts and floods occur yearly worldwide. Tornadoes cannot be stopped from forming; nor can their directions afterward be predicted or controlled. Lightning strikes during thunderstorms are unpredictable, as are volcanic eruptions, earthquakes, and tsunamis.

When it comes to climate, adapt, not control is the appropriate response. We've done it for ages. Adapting is likely to continue long into the future. When it rains, umbrellas and raincoats are the order of the day. In the cold of winter heavy clothing, jackets, caps, boots, and gloves, allow ventures outside. Evacuating people from the areas of anticipated volcanic eruptions or massive floods is usually the correct response. Relocating is sometimes appropriate. Respond by adapting.

Migrations have always been and will remain a popular response to climate change. 80,000 years ago human populations moved out of Africa, perhaps in response to climate, poor food supply, inadequate water, or any combination. It's probably true that poor climate led to water and/or food supply problems. Retirees in the northern U .S. A. flee the cold of winter for Florida or Arizona in the south. Tens of thousands of residents left New Orleans after Katrina.

Having said all of this about climate control, cleaning up the air, water, and land, while maintaining the ecosystem in balance, is essential. This is not climate control. It's housekeeping on a planetary scale. It happens after the fact; climate control, if ever possible, will come beforehand, preventing hurricanes, tornadoes, and tsunamis. No major cleanup will be required after prevention.

# Chapter 13

# *A FEW LOOSE ENDS*

## *SEA LEVEL RISE AND LOSS OF LIVING SPACE*

As coastlines change with rising sea levels, some living space will be (temporarily) lost. There will still be plenty of living space in most countries. The Texas vs. world population example supports this. Only for a few small or over populated low countries or islands will this be a problem. Since the changes will occur very slowly, they can anticipate and adapt. The 3 obvious options are population control (or reduction), migration, or rebuilding using multiple-story dwellings on higher ground. It is highly likely that aid will be available from a number of organizations and/or friendly well-to-do countries.

Actually, the overall land area in the world will increase. If all the ice melts, land in Greenland, Antarctica, and that covered by glaciers will become available. Antarctica alone is about twice as large as Australia.

## *WILL U.S. AGRICULTURAL LAND BECOME A DUST BOWL?*

One of the biggest fears often heard is that the U.S. prime agricultural area will turn into a dust bowl, unable to grow the grains as a result of GW. If it happens, there will be a lengthy transition. Though unsuitable for growth of grains, it may well be able to support sugar cane, citrus fruits, bananas, pineapples, and other fruits and vegetables for several generations.

If at last GW decimates U.S. agricultural land (9373 x 10*3 KM^2), other prime land not currently available will open. The frozen tundra in northern Canada and Siberia, not tilled for thousands of years, will become the world's breadbasket, perfect for growing grains that will not grow in the U.S then. There are 9776 x 10^3 KM^2 and more than 9853 x 10^3 KM*2 of land, respectively, there. That's more than twice what might be lost here.

## *TIPPING POINTS (TPS)*

I don't know what they are because I've never heard one defined. Maybe I haven't done enough archival research, but TP has been mentioned (without definition) by experts in several interviews I've seen. So I have inferred a definition from their words and attitudes. (Not having a definition except my own for what I'm talking about puts me on dangerous ground here.)

It seems that a TP in climate is defined by some set of parameters that give new very adverse living conditions. It can only get worse as future TPs are passed. And guess what; we can't do anything to get back to the wonderful climate, that used to please us so much. The one that generated a hurricane that devastated New Orleans, that created a tsunami which overwhelmed parts of Indonesia, that gave us Mt. St. Helens in 1980, that yearly created tornado alley in the U.S., that gave periodically increasing numbers of big worldwide floods since 1985 (Remember Chapter 9?), etc.

It has been implied that TPs originate in simulations. If true, their credibility evaporates. Common sense and a look back at some previous hard evidence should make us skeptical of TPs.

Review the graphs in Figure 2. At first glance they look gnarly having a bad case of the ups and downs. Yet there were 4 nice smooth warming periods with turning points (not TPs) at tops and bottoms. The LTCCs were hammered irregularly by megavolcanoes, but again and again afterward came the GWs. It conveys a sense or consistent alternating repetition. Each has a turning point (not a TP) at the top and bottom. Eventually parameters equilibrate. Not even a megavolcanic eruption produces a TP. So many events; so many turnarounds and repetitions; yet no TPs. The evidence and calculations presented up to this point indicate that mankind, if it continues operating with no all out nuclear war, while moving away from hydrocarbon energy sources, is not likely to create a TP in the next 200 years.

## *ABOUT CARBON DIOXIDE*

The composition of the atmosphere by volume is currently about 78.08% nitrogen, 20.95% oxygen, .036% (360 PPM) $CO_2$, and traces of a few others. A book by Nobel Prize winner, Linus Pauling, College Chemistry, copyright 1955 states it was 78.03% nitrogen, 20.99% oxygen, .03% (300 PPM) $CO_2$, plus others. That's an increase of 60 PPM in $CO_2$ concentration since 1955.

The percentage increase is usually stated as 20% - [60 PPM/300 PPM] x 100% = 20%. Hardly ever is it heard that it has increased from 300 PPM to 360 PPM. Never is it quoted as 036% vs..030% of the atmosphere. These numbers are too small, as opposed to a 20% increase, to convince the public that $CO_2$ is a threat via GW.

In the time of the dinosaurs the CO2 concentration was at least 6 to 10 times higher than now. Plant life flourished. Produce in supermarkets today is generally larger than it was in the 1950s in part due to higher CO2 levels. It has been proven that plants and their fruits grow stronger and larger in CO2-rich air. CO2 is food for plants. It's used to make every organic chemical they produce. They get it from the air, not the ground.

Above, the 400 PPM decrease in oxygen concentration was shown as a decrease from 20.99% to 20.95%. Loss of oxygen is far more significant than the increase in CO2. We require oxygen to live. It looks like our focus is on the wrong gas.

Burning hydrocarbons to get energy removes oxygen from the atmosphere. Deforestation eliminates the trees that replenish the air's oxygen. So the overheating via the greenhouse effect is not the primary reason to stop using hydrocarbons. Burning hydrocarbons combined with deforestation is the real threat to our future.

## THE AGRICULTURAL REVOLUTION

Of the 3 most current "revolutions" in human history, Agricultural, Industrial, and Digital, the longest is Agricultural – over 3000 years. It was then that we started "managing" the land for food production.

A major part of that management was clear-cutting of forests and burning timber. It's still an ongoing objectionable practice. Burning the world's forests is mainly responsible for the massive CO2 increase prior to 400 years ago. Since then the Industrial Revolution has been an additional major contributor.

Above is our contribution to changing the composition of the atmosphere, but not to significantly warming it. (See T^4 dependence below.)

## EARTH RADIATES HEAT TO SPACE

The rate at which heat is radiated from an object whose temperature is above absolute zero depends on its temperature in degrees Kelvin. One Kelvin degree is the same size as 1 Centigrade degree, but 0 degrees Centigrade is about 273 degrees Kelvin. Absolute 0, 0 degrees K, is about –273 C. 27 C (80.6 F) is 300 K. Adding 273 degrees to a Centigrade temperature gives the corresponding Kelvin temperature.

This rate is expressed by the equation:

$$Q = A \times T^4$$

A is a constant and T, the Kelvin temperature, is raised the 4th power. This is an extremely strong dependence showing that heat sheds extremely fast as temperature rises.

If the temperature of a body at 300 K (80.6 F) increases 3 Centigrade degrees to 303 K, that's a 1% temperature rise. Because of the 4th power relationship, Q increases by 4%. That's enormous protection against overheating. (Try raising 1 to 1.01 – a 1% increase - with a regular calculator. Then multiply 1.01 by itself 4 times – 1.01 x 1.01 x 1.01 x 1.01 gives about 1.04+; a 4%+ increase.)

This T^4 dependence makes overheating very unlikely. Though the green-house effect retains some heat, it is unlikely to get out of hand. A possible example of this control is clear-cutting and burning of forests during the Agricultural Revolution and overlapping Industrial Revolution. Apparently the tremendous amount of heat generated during 3000 years of burning trees, coal, and hydrocarbons was easily shed by planet Earth. It is probable that some of that energy melted ice and warmed Earth's oceans, but the T^4 radiation was and is always functioning to limit temperature rise.

# Chapter 14

# CONCLUSION

When I began this work I thought that our energy contribution to the climate was insignificant. I felt that 2 immense furnaces, the sun and the Earth's core, were the primary controllers of climate. As I crunched the numbers related to the oceans and the ice, calculations confirmed my view.

Then I found the climatic fingerprints in Figure 2. They show that GW is a natural event. They display how our planet manages its cool down at this time in its super-long-term cooling cycle. (Remember that several billion years ago Earth was too hot to support life, and several billion years from now it will be far too cold for that.)

Movement of the tectonic plates and subduction of the sea floor under continents produce most of the heat and the $CO_2$ that are present during GWs. Heat and $CO_2$ are also siphoned from the interior via volcanoes. Increasing concentrations of $CO_2$ are the evidence that GW may be occurring; $CO_2$ is not the cause of GW via the greenhouse effect.

The GW ice-age-100,000 year cycle can now be better understood in terms of Earth bound parameters. At the end of an ice age polar caps are melting. Ice bridges between the ice caps and continents are broken. The tectonic plates begin to move more freely. Subduction of the sea floor under the continents generates heat and $CO_2$ throughout the GW period. Volcanoes spew both from the interior into our environment.

The resulting planet-wide temperature rise gives increasing evaporation of Earth's water. More water vapor in the air leads to more precipitation. There comes a point where excessive precipitation begins to give massive frequent floods worldwide.  Flooding and the size of floods increase with time. This leads to more evaporative cooling and growing inertia on the continents, as water accumulates on and inside exposed crust.

Increased continental inertia via water accumulation slows subduction. This reduces heating and $CO_2$ production. The result is a GW turnaround. Afterward these conditions continue, giving falling temperatures and lower $CO_2$ concentrations.

Decreasing temperatures turn rain into snow. As it piles up over a long period of time, it compresses the snow beneath it into ice. The polar caps grow. Ice bridges between both polar caps and 6 continents lock them together creating massive inertia to tectonic plate movement.

Over tens of thousands of years massive ice and snow accumulation on the continents compress the crust. This thinner crust has a much higher thermal conductivity. The temperature at the crust-ice interface has been lowered by more that 20 Fahrenheit degrees vs. GW periods. Reduced temperature, increased thermal conductivity, and thinned crust, lead to greater heat flow to the ice-crust interface. The huge layer of insulating snow above trap most of the heat. It begins melting the ice at the interface leaving the temperature there unchanged. Water, forced out by extreme pressure, erodes the ice.

At some point the heat arriving at the ice perimeter is large enough that it melts snow there faster than it can fall. The perimeter begins to decrease. More and more compressed crust becomes exposed. The melt down has begun.

This melt down near the end of an ice age continues, until the ice bridges are broken, and the tectonic plates are free to move again. In the examples in Figure 2 the 4 GW cycles never reach this point. Massive volcanic eruptions (some megavolcanic) are so frequent during ice ages, that near the end one or more usually occurs breaking the ice bridges and freeing the tectonic plates. This completes the cycle, and GW begins.

There is a way to make sure that we will not contribute to warming our environment. Extract kinetic energy (KE) from our environment, use it as is required, and return it to the environment, usually in the form of heat. (Sounds like a contradiction, doesn't it? Read on.)

KE is omnipresent in vast quantities in the form of wind, waves, underwater ocean currents, and tides. Wind farms, wave energy extraction machines, and tidal turbines are already being used to generate electricity, which can be used for all of civilization's energy needs.

The net exchange of energy is zero without endangering the environment. Pollutants, if any, are negligible, and the heat returned creates temperature and pressure differentials to make more air movement – wind. The returned heat changes to KE, which can be harvested again.

Currently latent energy in coal, oil, gas, nuclear, etc. is released from storage and added to our environment. These warm and pollute. The process of switching over is now ongoing.

Electricity produced by alternative energy sources can be used to generate hydrogen (Currently experimental in Norway). It serves as a storage mechanism not too different in concept from the storing oil in huge tanks. Hydrogen can be burned to drive generators to make electricity and to heat. The only 2 byproducts are energy and water.

I hope that this text aids thinking about GW and making calculations. With all the hype and misinformation out there it is necessary to be able to analyze it. Use common sense; use a simple calculator to crunch the numbers. Most of all think. Herein are provided many of the statistics and conversion factors needed.

# PROJECTED 20TH CENTURY WORLD ENERGY USE IN mQ

This table was generated starting with 20.51 mQ used in the year 1900. The growth rate is 3%. A complete table could not be found. Beginning and ending values are very close to real ones.

| YEAR | PERIODICALLY | CUMULATIVE |
|------|--------------|------------|
| 1901 | 21 | 21 |
| 1904 | 23 | 88 |
| 1908 | 26 | 188 |
| 1912 | 29 | 300 |
| 1916 | 33 | 426 |
| 1920 | 37 | 568 |
| 1924 | 42 | 727 |
| 1928 | 47 | 907 |
| 1932 | 53 | 1109 |
| 1936 | 59 | 1337 |
| 1940 | 67 | 1593 |
| 1944 | 75 | 1881 |
| 1948 | 85 | 2206 |
| 1952 | 95 | 2571 |
| 1956 | 107 | 2982 |
| 1960 | 121 | 3445 |
| 1964 | 136 | 3965 |
| 1968 | 153 | 4551 |
| 1972 | 172 | 5211 |
| 1976 | 194 | 5953 |
| 1980 | 218 | 6789 |
| 1984 | 246 | 7729 |
| 1988 | 276 | 8788 |
| 1992 | 311 | 9979 |
| 1996 | 350 | 11320 |
| 2000 | 394 | 12829 |

TOTAL ENERGY USED IN THE 20TH CENTURY IN mQ = 12,829

# PROJECTED 21ST CENTURY WORLD ENERGY USE IN mQ

This table was generated starting with 394 mQ used in the year 2000. The growth rate is 3%. Cumulative values are good through the year indicated on the left.

| YEAR | PERIODICALLY | CUMULATIVE |
|------|--------------|------------|
| 2001 | 406 | 406 |
| 2004 | 443 | 1698 |
| 2008 | 499 | 3609 |
| 2012 | 562 | 5759 |
| 2016 | 632 | 8180 |
| 2020 | 712 | 10905 |
| 2024 | 801 | 13971 |
| 2028 | 901 | 17422 |
| 2032 | 1015 | 21307 |
| 2036 | 1142 | 25679 |
| 2040 | 1285 | 30599 |
| 2044 | 1447 | 36138 |
| 2048 | 1628 | 42371 |
| 2052 | 1832 | 49387 |
| 2056 | 2062 | 57283 |
| 2060 | 2321 | 66170 |
| 2064 | 2613 | 76173 |
| 2068 | 2941 | 87431 |
| 2072 | 3310 | 100102 |
| 2076 | 3725 | 114364 |
| 2080 | 4193 | 130415 |
| 2084 | 4719 | 148481 |
| 2088 | 5311 | 168815 |
| 2092 | 5977 | 191700 |
| 2096 | 6728 | 217458 |
| 2100 | 7572 | 246449 |

TOTAL USE PROJECTED IN 21ST CENTURY IN mQ = 246,449

# PROJECTED 22ND CENTURY WORLD ENERGY USE IN mQ

This table was generated starting with 7572 mQ used in the year 2100. The growth rate is 3%. Cumulative values are good through the year indicated on the left.

| YEAR | PERIODICALLY | CUMULATIVE |
|------|--------------|------------|
| 2101 | 7799 | 7799 |
| 2104 | 8522 | 32629 |
| 2108 | 9592 | 69353 |
| 2112 | 10796 | 110686 |
| 2116 | 12151 | 157207 |
| 2120 | 13676 | 209566 |
| 2124 | 15392 | 268497 |
| 2128 | 17324 | 334825 |
| 2132 | 19499 | 409477 |
| 2136 | 21946 | 493499 |
| 2140 | 24700 | 588066 |
| 2144 | 27800 | 694502 |
| 2148 | 31289 | 814297 |
| 2152 | 35216 | 949127 |
| 2156 | 39636 | 1100880 |
| 2160 | 44611 | 1271678 |
| 2164 | 50210 | 1463914 |
| 2168 | 56512 | 1680276 |
| 2172 | 63605 | 1923794 |
| 2176 | 71588 | 2197876 |
| 2180 | 80573 | 2506357 |
| 2184 | 90685 | 2853555 |
| 2188 | 102067 | 3244330 |
| 2192 | 114877 | 3684150 |
| 2196 | 129295 | 4179171 |
| 2200 | 145523 | 4736323 |

TOTAL USE PROJECTED IN THE 22ND CENTURY IN mQ = 4,736,323

# *HEAT CAPACITY OF EARTH'S AIR*

Specific heat of air  = 1035  JOULES/KG at constant pressure

Mass of the Earth's atmosphere =  $5.14 \times 10^{18}$ KG

Energy = (SPECIFIC HEAT OF AIR) x (MASS OF AIR)
Energy =      1035 JOULES/KG   x    $5.14 \times 10^{18}$ KG

Energy required to raise the temperature of the Earth's atmosphere 1 Centigrade degree =     $5320 \times 10^{18}$  JOULES

Energy in milliquads  =    5320 mQ

# *HEAT CAPACITY OF EARTH'S WATER (OCEANS)*

Specific heat of water   4180  JOULES/KG

Mass of the Earth's water (oceans)  1.36 x 10^21 KG

Energy = (SPECIFIC HEAT OF WATER) x (MASS OF WATER)
Energy =        4180 JOULES/KG        x   1.36 x 10^21 KG

Energy required to raise the temperature of the Earth's  oceans 1 Centigrade degree    5.684801 x 10^24 JOULES

Energy in milliquads  5,684,801 mQ

# *ENERGY NEEDED TO MELT ALL OF EARTH'S ICE*

It is estimated that there are 33 million KM^3 (cubic kilometers) of ice on our planet. When ice melts it contracts about 10%.

Volume of water = .9 X 33,000,000 KM^3 = about 29,700,000 KM^3 of water. (rounding to 30,000,000 KM^3 of melt water)

<u>The mass of the ice is still the same as the mass of the water.</u>

Now converting KM^3 to M^3 (cubic meters):

30,000,000 KM^3 water X [ <u>1000 Meters</u> ]^3  =  3 X 10^16 M^3
                               KM^3

Then converting M^3 to CM^3 or CC or ML (cubic centimeters or millimeters):

3 X 10^16 M^3  X   [ <u>100 cm</u> ]^3  gives 3 X 10^22 CC of water
                      1 M^3

1 CC of water has about 1 gram (GM) mass.

Convert CC of water to grams of water.

3 X 10^22 CC of water X  <u>1 GM pf water</u> = 3 X 10^22 GM of water
                          1 CC of water

Converting to kilograms (KG):

3 X 10^22 GM of water x  <u>1 KG water</u>   =  3 X 10^19 KG water
                      1000 KG water

Since 1 KG of water = 1 KG of ice, 3 X 10^19 KG of water = 3 X 10^19 KG of ice.

To melt 1 KG of ice it takes 334,000 J (joules); to freeze 1 KG of water 334,000 J must be removed. That amount of energy is called the heat of fusion of water. Now the total mass of ice can be multiplied by the heat of fusion to get the total energy required to melt all Earth's ice.

$$3 \times 10^{*}19 \text{ KG total ice mass} \times 334,000 \text{ J /KG} = 10^{25} \text{ J}$$

Converting to mQ (milliquads) gives:

$$10^{25} \text{ J} \times \frac{1 \text{ mQ}}{10^{18} \text{ J}} = 10^{7} \text{ mQ}$$

It takes 10 million mQ to melt all the ice on earth.

# *ICE MELT LEADS TO SEA LEVEL RISE*

Area of the Earth:   $5.1 \times 10^8$ KM^2

Area of the oceans:  $3.61 \times 10^8$ KM^2

Volume of ice melt water:   $3 \times 10^7$ KM^3

90% of this water comes from Greenland and Antarctica.

$.9 \times 3 \times 10^7$ KM^3 = $2.7 \times 10^7$ KM^3

Since volume = area x height, height = volume/area.

Dividing:     $2.7 \times 10^7$ KM^3/ ($3.61 \times 10^8$ KM^2)  =  .0748 KM

---

Converting to meters gives: .0748 KM x 1000 M/KM   =  74.8 Meters

---

Since some of this melt water is below sea level it will not raise sea level.  Assuming 10% to 25% is below sea level, a more accurate range is:

---

56.1 to 67.3 Meters Rise

---

In this calculation I have ignored the 5% of water from glacial melting.

# *ENERGY STORED IN THE EARTH'S CORE*

| Assumed Core Temp in Deg.C | Core Temp – Ambient (Amb. = 15 Deg. C) | Energy in Joules | Energy (mQ) |
|---|---|---|---|
| 5000 | 4985 | 3.62 x 10^30 | 3.62 x 10^12 |
| 5500 | 5485 | 3.98 x 10^30 | 3.98 x 10^12 |
| 6000 | 5985 | 4.35 x 10^30 | 4.35 x 10^12 |
| 6500 | 6485 | 4.71 x 10^30 | 4.71 x 10^12 |
| 7000 | 6585 | 5.07 x 10^30 | 5.07 x 10^12 |

SOME CONSTANTS USED IN THESE CALCULATIONS

RE = 6371 KM          RADIUS OF THE EARTM
RS = 1220 KM          RADIUS OF THE SOLID METAL CORE
RO = 3400 KM          OUTER RADIUS OF THE LIQUID CORE

VS = 7.69*10^18 M^3   VOLUME OF THE SOLID SPERICAL CORE
VO = 1.65*10^20 M^3   VOLUME OF THE OUTER SPHERE OF
                      SOLID + LIQUID
VL = 1.57*10^20 M^3   VOLUME OF LIQUID = VO-VS

SH = 440 JOULES/KG; SPECIFIC HEAT OF IRON IN J/KG
(FOR BOTH THE SOLID AND LIQUID CORE)

HEAT FROM SOLIDIFYING THE LIQUID IRON = 2.14 x 10^10 mQ
HEAT OF FUSION OF IRON = 13810 J/KG

Energy gained by removing heat from the outer liquid portion of the core and solidifying it is 2 orders of magnitude lower than that gained by cooling the cores down to ambient; it can be ignored here. That is 2.14 x 10*10 mQ is much less than say 3.62 x 10^12 mQ and can be ignored when compared to the thermal energy in the right-most column of the table above.

# *ENERGY STORED IN THE EARTH'S 2 MANTELS*

In making this calculation the upper and lower mantels were treated as 10 shells of equal mass having the same density and specific heat. The difference between the core temperature and ambient on the surface was divided by 10. At each radius the temperature was incremented downward by this difference. The temperature assigned to each shell was that at its outer radius minus ½ the difference.

| SHELL NUMBER (#1 IS CLOSEST TO CORE) | OUTER (KM) | RADIUS (MILES) |
|---|---|---|
| 1 | 943 | 2450 |
| 2 | 4368 | 2714 |
| 3 | 4723 | 2935 |
| 4 | 5032 | 3127 |
| 5 | 5306 | 3297 |
| 6 | 3452 | 5555 |
| 7 | 5783 | 3594 |
| 8 | 5995 | 3725 |
| 9 | 6193 | 3848 |
| 10 | 6378 | 3964 |

| CORE TEMP DEG. C | 10th SHELL CENTER TEMP. | TOTAL ENERGY JOULES | mQ |
|---|---|---|---|
| 5000 | 264.25 | 8.960376E+30 | 8.960376E+12 |
| 5500 | 289.25 | 9.859108E+30 | 9.859108E+12 |
| 6000 | 314.25 | 1.075784E+31 | 1.075784E+13 |
| 6500 | 339.25 | 1.165658E+31 | 1.165658E+13 |
| 7000 | 364.25 | 1.255531E+31 | 1.255531E+13 |

# DARTMOUTH FLOOD OBSERVATORY
# ARCHIVES    1985-2007

| YEAR | # OF LARGE FLOODS | YEAR | # OF LARGE FLOODS |
|------|-------------------|------|-------------------|
| 1985 | 69 | 1996 | 98 |
| 1986 | 70 | 1897 | 210 |
| 1987 | 45 | 1998 | 184 |
| 1988 | 111 | 1999 | 101 |
| 1989 | 150 | 2000 | 102 |
| 1990 | 104 | 2001 | 171 |
| 1991 | 123 | 2002 | 261 |
| 1992 | 115 | 2003 | 298 |
| 1993 | 99 | 2004 | 194 |
| 1994 | 121 | 2005 | 171 |
| 1995 | 112 | 2006 | 232 |
|      |     | 2007 | 244 |

The Dartmouth Flood Observatory
Web site at:  http: // www.darmouth. edu/%7Efloods/Archives/index.htm
has great worldwide maps showing annual flooding for each year
from 1985 through 2007 as well as the annual numbers shown here.

APPENDIX K

# FIGURES 5    4 GW SLOPES ( INVERSES)

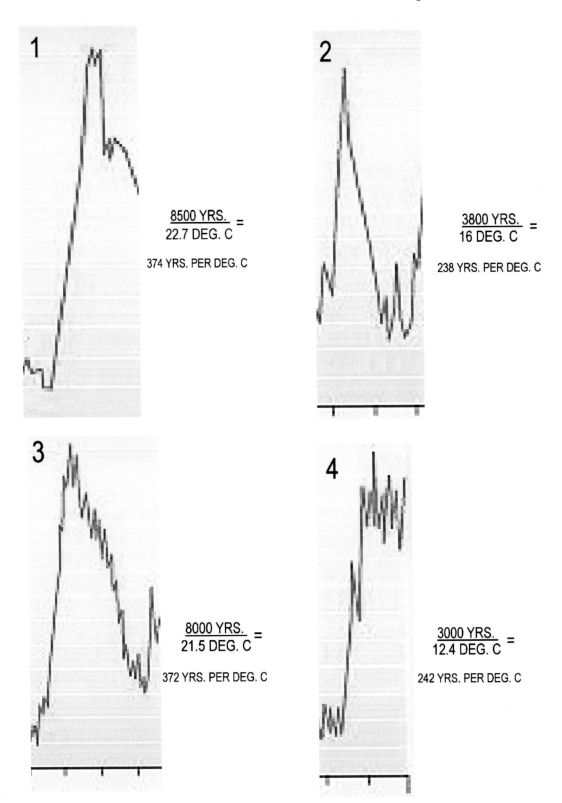

1

$$\frac{8500 \text{ YRS.}}{22.7 \text{ DEG. C}} =$$

374 YRS. PER DEG. C

2

$$\frac{3800 \text{ YRS.}}{16 \text{ DEG. C}} =$$

238 YRS. PER DEG. C

3

$$\frac{8000 \text{ YRS.}}{21.5 \text{ DEG. C}} =$$

372 YRS. PER DEG. C

4

$$\frac{3000 \text{ YRS.}}{12.4 \text{ DEG. C}} =$$

242 YRS. PER DEG. C

# ABOUT THE AUTHOR

Mr. George Sourlis attended Purdue University from 1960 through 1964 when he received his BS. He holds a double major in physics and math. In 1965 he entered the Master degree program at the University of Arkansas majoring in physics. After 2 years of course and thesis work, he took a position teaching physics at Drury College in Springfield, Missouri in 1967 112 miles from the U. of A.

During his teaching there he commuted on weekends and spent his summers at the U. of A. finishing his thesis work. In 1969 with that completed he had earned an M.S. in physics.

In December of 1969 he left academic work to find something doing research. He was hired to work in the Applied Research Department at Videojet Systems Inc., a division of the A.B. Dick Company at that time. In his 14 years there he studied and help develop a continuous-jet industrial ink jet marking machine, which is now used world wide to bar code mail and date code food and beverage product containers. The public sees these codes daily on their U.S. mail, bottoms of beverage cans, and on other grocery products. They are printed in a dot matrix format as bar codes or alpha-numerics

During his time doing applied research as a physicist for Videojet Systems, Inc. he patented 1 item and developed 1 proprietary process. He is an expert in the process of continuous-jet ink jet printing.

The company changed hands in the early 1980s, and in 1987 he left to start his own company in an unrelated field. He retired in 2002.

Printed in the United States
By Bookmasters